Axolotls

Ruby Daniels

Big Buddy Books

An Imprint of Abdo Publishing
abdobooks.com

abdobooks.com

Published by Abdo Publishing, a division of ABDO, PO Box 398166, Minneapolis, Minnesota 55439.

Printed in the United States of America, North Mankato, Minnesota
102024
012025

Design: Elena Klinkner, Mighty Media, Inc.
Production: Mighty Media, Inc.
Editor: Liz Salzmann
Cover Photograph: axolotlowner/Shutterstock Images
Interior Photographs: Aleron Val/Shutterstock Images, p. 15 (middle); axolotlowner/Shutterstock Images, p. 13; Carter Charles Johnson/Shutterstock Images, p. 15 (top); cherokee4/Adobe Stock, p. 5; Eduardo/Adobe Stock, p. 21; Eyepix/NurPhoto/Associated Press, p. 15 (bottom); Francisco Gomez Sosa/Shutterstock Images, p. 17; Gobierno CDMX/Flickr, p. 14 (bottom); Guillermo Guerao Serra/Shutterstock Images, pp. 9, 11; Micha/Adobe Stock, p. 7; Real Window Creative/Shutterstock Images, p. 25; SingerGM/Shutterstock Images, p. 27; SiPakne/Shutterstock Images, p. 23; stas111/Adobe Stock, p. 18 (compass rose); tanarch/Adobe Stock, pp. 18–19 (maps); Ива Димова/Adobe Stock, p. 29
Design Elements: flovie/Shutterstock Images (patchwork pattern); Mighty Media, Inc. (series logos & icons)

Library of Congress Control Number: 2024938320

Publisher's Cataloging-in-Publication Data
Names: Daniels, Ruby, author.
Title: Axolotls / by Ruby Daniels
Description: Minneapolis, Minnesota : ABDO Publishing, 2025 | Series: Odd but adorable animals | Includes online resources and index.
Identifiers: ISBN 9781098295103 (lib. bdg.) | ISBN 9798384915157 (ebook)
Subjects: LCSH: Axolotls--Juvenile literature. | Amphibians--Juvenile literature. | Salamanders--Juvenile literature. | Freshwater animals--Juvenile literature. | Aquatic animals--Juvenile literature. | Curiosities and wonders--Juvenile literature.
Classification: DDC 597.85--dc23

Contents

An Axolotl Encounter 4
A Closer Look 6
Life as an Axolotl 8
Strange Salamanders 10
Creature Feature 14
Disappearing Homes 16
Location Station 18
Critter Culture 20
Amazing Investigations 24
Threats and Hope. 26
Odd or Adorable?. 28
Glossary. 30
Online Resources 31
Index 32

An Axolotl Encounter

In Mexico City, Mexico, you take a boat ride on Lake Xochimilco (So-chee-MIL-co). Suddenly, something small and pink appears at the surface. The creature looks up, and you see a face with frills along the sides. It almost seems to smile at you. You've just spotted the odd but adorable axolotl!

The wide, slightly upturned shape of an axolotl's mouth makes it appear to smile.

A Closer Look

Most wild axolotls are grayish brown in color. But their colors can vary. Pet axolotls are often bred to have different skin colors. Leucistic axolotls are popular for their pale pinkish skin. Adult axolotls can be 6 to 18 inches (15 to 46 cm) long. They weigh 2 to 8 ounces (57 to 227 g).

The different axolotl colors are called morphs. Common morphs include (*left to right*) melanoid, leucistic, and golden albino.

Life as an Axolotl

Axolotls are a type of salamander. Like other salamanders, they lay eggs. Most salamanders live at least partly on land. But axolotls live their whole lives underwater. They have fins and webbed feet to help them swim.

Axolotls usually hunt at night. They eat snails, insects, and small fish.

Strange Salamanders

Like most other salamanders, axolotls lay eggs that hatch in the water. The babies are called larvae. Most salamander larvae go through **metamorphosis** and move onto land. But axolotls keep many of their baby features so they can stay underwater. These include frilly gills on the sides of their faces. The gills allow axolotls to breathe underwater.

Most salamanders have eyelids, but axolotls don't.

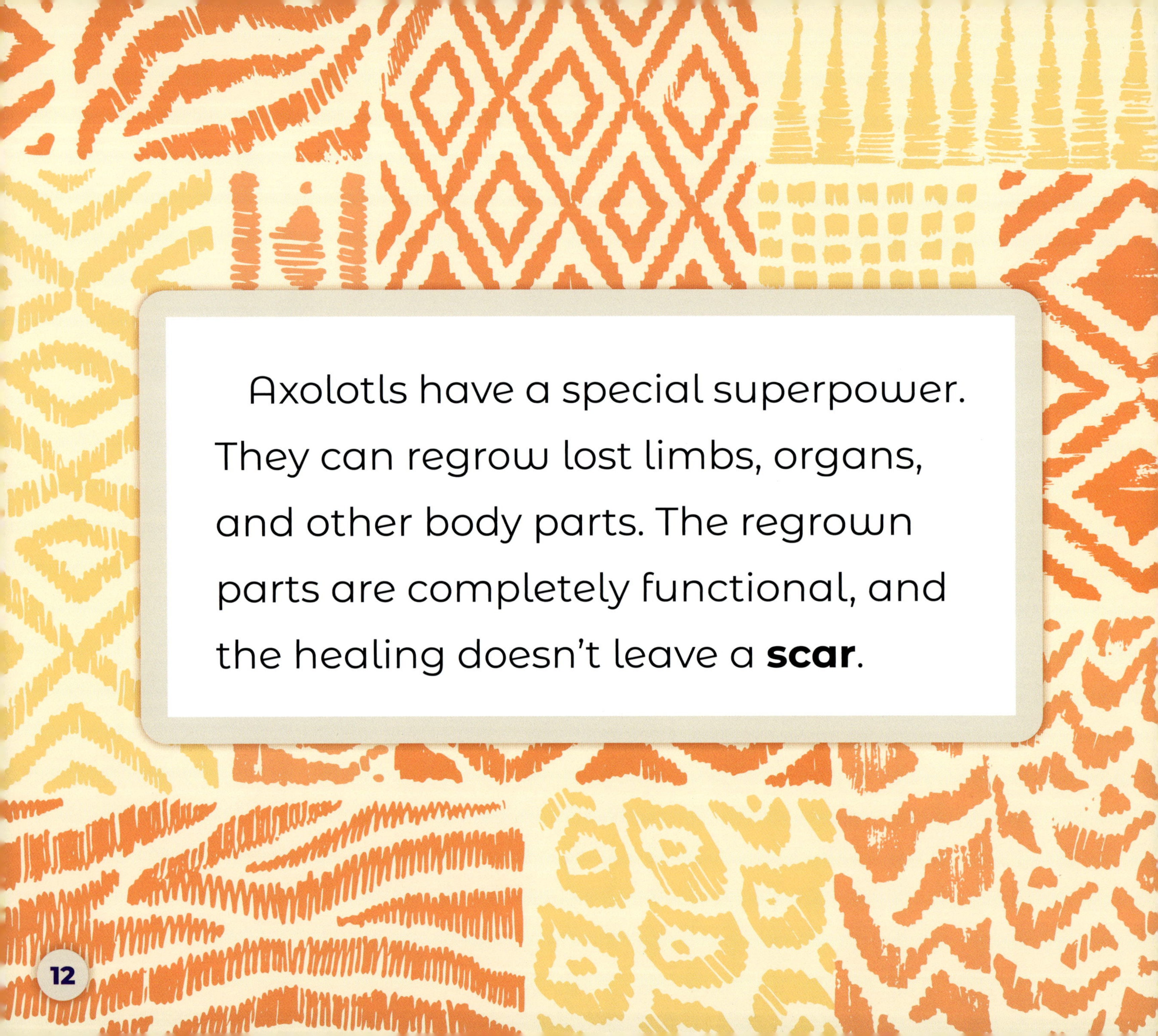

Axolotls have a special superpower. They can regrow lost limbs, organs, and other body parts. The regrown parts are completely functional, and the healing doesn't leave a **scar**.

An axolotl's regrown limb may not look exactly like its other limbs.

Creature Feature

The word *axolotl* is from an ancient **Aztec** language. It can mean "water dog" or "water monster."

Mexico City, Mexico, held a contest to create a set of **emoji** that would officially represent the city. The winner was a set of axolotl emoji!

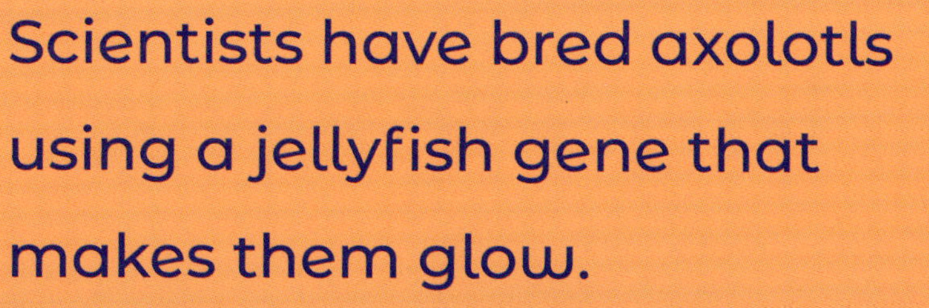

Scientists have bred axolotls using a jellyfish gene that makes them glow.

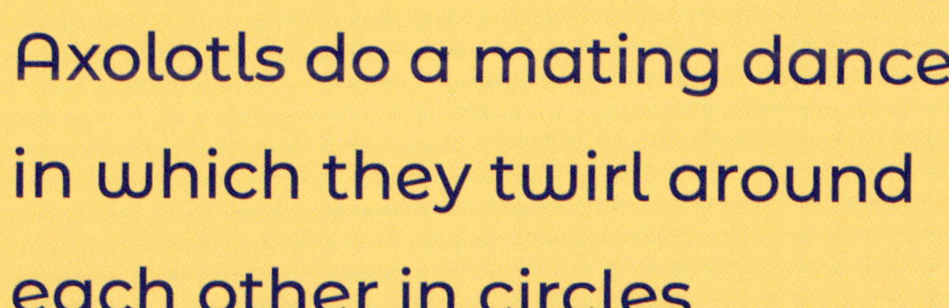

Axolotls do a mating dance in which they twirl around each other in circles.

Female axolotls lay hundreds of eggs at a time. From the moment they hatch, axolotls live on their own.

Disappearing Homes

Wild axolotls have lost much of their native **habitat** due to human activity. This includes polluting the water and **draining** lakes. Today, axolotls naturally exist only in **wetlands** in and near Mexico City, Mexico. However, axolotls are kept as pets all over the world! Whether living in the wild or with humans, axolotls need fresh water and small creatures to eat.

Lake Xochimilco in Mexico City is one area where axolotls live. It is now a UNESCO World Heritage Site.

Location Station

Axolotl Habitat
ATLANTIC OCEAN
BELIZE
GUATEMALA
ARCTIC OCEAN
NORTH AMERICA
United States
EUROPE
ASIA
ATLANTIC OCEAN
PACIFIC OCEAN
PACIFIC OCEAN
AFRICA
SOUTH AMERICA
INDIAN OCEAN
ATLANTIC OCEAN
AUSTRALIA
SOUTHERN OCEAN

Critter Culture

Since ancient times, axolotls have been an important part of Mexican **culture**. There is an ancient **Aztec** myth about a god named Xolotl. According to the myth, he turned himself into a salamander to hide. Today, axolotls are included in public art around Mexico City.

In 2021, Mexico honored the axolotl's importance to Mexican history and culture by putting it on the 50-peso bill.

Axolotls have also affected **cultures** around the world. The axolotl pet trade began in 1864, when European explorers brought them back from Mexico. Demand for pet axolotls has grown recently due to their appearance on social media and in video games.

However, axolotls can be hard to care for properly. Many places have laws to prevent people from setting them free in unsuitable **habitats**.

Games such as *Minecraft* (*pictured*) and *Fortnite* include axolotls.

Amazing Investigations

Many scientists study axolotls. In addition to being able to regrow body parts, axolotls also rarely get **cancer**. Scientists are trying to learn how axolotls resist cancer. They think this could lead to new **treatments** for cancer in human patients.

One of the main laboratories where axolotls are studied is at the University of Kentucky in Lexington.

Threats and Hope

Axolotls are considered **critically endangered**. Human activities such as farming and construction have destroyed their **habitats**. Experts say there may be as few as 100 axolotls left in the wild.

However, some people are working to save axolotls. For example, they build **filters** to keep pollution out of axolotl habitats. One day, axolotls may **thrive** again in the wild!

People have introduced non-native fish such as tilapia into areas where axolotls live. These fish eat axolotl eggs and compete with axolotls for food.

Odd or Adorable?

Axolotls are unusual and beloved animals. They look strange, but they're also considered cute by many people around the world. What do you think makes an animal odd? What makes an animal adorable? Do you think axolotls are odd, adorable, or both? Why?

Axolotls are especially popular pets in China and Japan.

Glossary

Aztec—a people who ruled a large empire in present-day Mexico in the 1400s and early 1500s.

cancer—any of a group of very harmful diseases that cause a body's cells to become unhealthy.

critically endangered—having almost none left in the world.

culture (KUHL-chuhr)—the arts, beliefs, and ways of life of a group of people.

drain—to remove liquid from something.

emoji—a small symbol or picture that can be typed in an email, a text, or an online post.

filter—a material with tiny openings through which liquid passes. When liquid passes through the filter, it catches objects and separates them.

habitat—a place where a living thing is naturally found.

metamorphosis (meh-tuh-MAWR-fuh-suhs)—changes that take place when some insects and salamanders change from larvae to adults.

scar—a mark that stays after a wound heals.

thrive—to be healthy and grow.

treatment—medical care for an illness or injury.

wetland—land mostly covered by water.

Online Resources

To learn more about axolotls, please visit **abdobooklinks.com** or scan this QR code. These links are routinely monitored and updated to provide the most current information available.

Index

art, 20
axolotl (word), 14
Aztec, 14, 20
breeding, 6, 15
cancer, 24
culture, 20, 21, 22
eggs, 8, 10, 15, 27
emoji, 14
endangerment, 16, 22, 26, 27
features, 4, 5, 6, 7, 8, 10, 12, 24
food, 9, 16
gills, 4, 10
habitats, 16, 17, 18, 19, 26
larvae, 10
laws, 22
length, 6
map, 18, 19
mating dances, 15
Mexico City, Mexico, 4, 14, 16, 17, 18, 20
pets, 6, 16, 22, 29
regrowing, 12, 13, 24
salamanders, 8, 10, 11, 20
skin colors, 4, 6, 7
social media, 22
studies, 24, 25
UNESCO, 17
video games, 22, 23
water, 8, 10, 14, 16, 17
weight, 6
Xochimilco, Lake, 4, 17
Xolotl, 20